Men on the
MOON

A Tribute to Space Exploration
THE MISSIONS • THE MEN • THE LEGACY

Contents

Published by World Publications Group, Inc.
140 Laurel Street
East Bridgewater, MA 02333
www.wrldpub.com

© Instinctive Product Development 2013

Packaged by Instinctive Product Development for World Publications Group, Inc.

Printed in China

ISBN: 978-1-4643-0184-1

Designed by: BrainWave

Creative Director: Kevin Gardner

Written by: Colin Salter, Helen Akitt, and Michael Heatley

Images courtesy of PA Photos and Wiki Commons

Introduction

■ **ABOVE:** Space Shuttle *Endeavour*, mounted on NASA's shuttle carrier aircraft, passes over the Golden Gate Bridge in San Francisco, 2012. *Endeavour* was making a final trek across the country to the California Science Center in Los Angeles, where it will be permanently displayed.

The space race that began as a battle of two superpowers was won by the United States when Neil Armstrong walked on the Moon in 1969. Or was it?

In fact, the Soviet Union was first to put a human being in space, and the two competitors finally came together in 1998 to found an International Space Station. By then, nearly three decades had passed since man last walked on the Moon. Yet our obsession with space exploration, fueled by *Star Trek* and many other science-fiction books, movies, and TV shows, remains as strong as ever.

This publication seeks to explain this fascination by charting man's attempted conquest of space and what it has left as its legacy. It lists all the significant dates, from the first unmanned Sputniks to the recent Mars lander, and lists some of the most notable space men (and women) who risked all for the cause. We particularly focus on Neil Armstrong, the man whose small step onto the Moon's surface was "one giant leap for mankind" and who passed away in 2012.

Many technological advances have emerged from the space program, spin-offs from humankind's duty to explore. Such staples of everyday life, from GPS systems to cordless power tools, can be traced back to this source. Yet it's the thrill of space exploration that continues to exert a strange fascination, and that state of affairs seems likely to continue. Whether or not man seeks "to boldly go where no man has gone before," the achievements of the past half-century are there to be examined and celebrated.

Space is still the final frontier, and every night when we see the Moon, we dream. This book is dedicated to that dream and all who have acted on it.

The Neil Armstrong Story
– A Tribute

Neil Alden Armstrong was born on August 5, 1930 on his grandparents' farm at Wapakoneta, Auglaize County, Ohio. He was destined for flight from the outset. He saw his first airplane at the age of two when his father took him to the Cleveland Air Races, and took his first flight four years later in the company of his father, on board a Tri-Motor – the Ford Motor Company's three-engine transport aircraft nicknamed the Tin Goose.

Young Neil was hooked, and took jobs working at the county airport to earn the money for flying lessons. He won his flight certificate before he got his driver's license, at the age of just 15 and, two years later, enroled at Purdue University, Indiana, to study for a degree in aeronautical engineering. It was a measure of Armstrong's natural aptitude for the subject that he was also offered a place at the Massachusetts Institute of Technology – but he turned it down on the advice of a friend, an MIT engineering graduate, who told him he didn't have to go that far to get a good education.

His studies were funded by a scholarship called the Holloway Plan, launched after World War II to provide new ways to enter the US Navy. Armstrong would study for two years, serve for three, and then return to university for two final

■ ABOVE: A youthful Neil Armstrong.

■ ABOVE: **Grumman Panther jets were some of the earliest aircraft Neil Armstrong flew.**

years. Naturally Armstrong's service, from 1949 to 1952, was as a naval aviator. By the time Neil returned to complete his degree he had flown 78 missions over Korea in Grumman Panther jets from USS *Essex*.

One of his earliest Korean sorties was nearly his last. Making a low-level bombing run in September 1951 he was hit by enemy anti-aircraft fire, lost control of his plane, and struck a tall post sticking out of the ground that shaved off about three feet of his starboard wing. Only his instinctive feel for his aircraft enabled him to fly it back into territory over which he could safely eject.

In 1955 he signed on as a civilian test pilot with NASA's forerunner, the National Advisory Committee for Aeronautics (NACA) and, for the next seven years, was stationed at Edwards Air Force Base in California. At first he flew observation missions in chase planes alongside experimental aircraft, but soon graduated to piloting the

experimental vehicles themselves.

From 1957 he was closely involved in the program to develop a rocket plane, which began with the Bell X-1. He piloted the North American X-15, a type that still holds the world record for the fastest speed by a manned aircraft, on seven occasions. Under Armstrong's control it once reached a speed of 3,989 mph – at Mach 5.74 the sixth fastest flight it ever made.

During one flight in 1962 he climbed to a height of 207,500 feet. On his descent he misjudged the angle of re-entry and, with the X-15's nose cone pitched too high, he bounced back off the atmosphere. The shallower descent he was forced to make as a result saw him overshoot the Edwards landing strip at 100,000 feet and at a speed of Mach 3 (around 2,000 mph). By the time he managed to turn and land the aircraft he had completed the longest X-15 flight on record.

Armstrong was a popular and highly respected member of the

■ BELOW: **This first official picture of *Sputnik 1* was issued in Moscow on October 9, 1957, showing the four-antennaed baby moon resting on a three-legged pedestal. Pravda described the satellite as one foot, 11 inches and gave its weight as 184 pounds. The outer hull is polished aluminum covered with a special protective material. Nitrogen gas is sealed inside.**

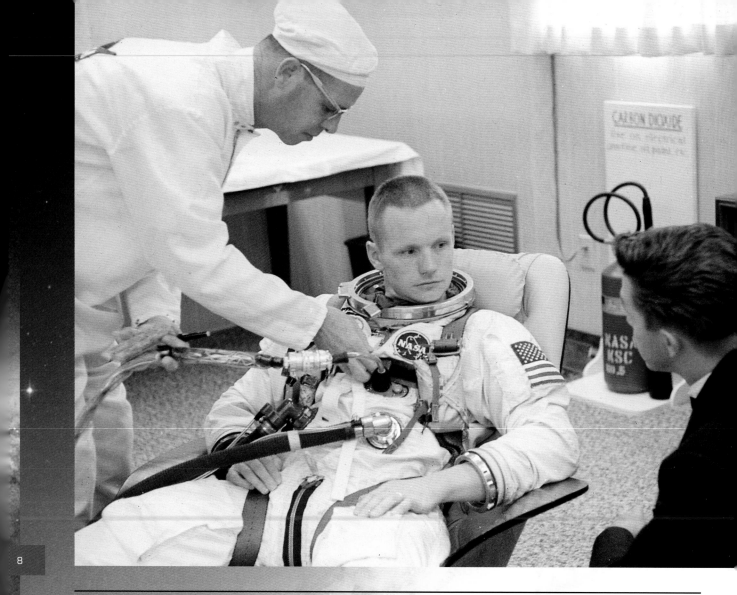

■ ABOVE: **Neil Armstrong during a suiting-up exercise March 9, 1966 at Cape Kennedy, Florida, in preparation for the** *Gemini 8* **flight.**

Edwards team. Although well regarded as a pilot, it was his innate understanding of aeronautical engineering that made him stand out from his peers. He had a feel for the fabric and construction of a plane, and this helped him predict how it would perform.

There may have been better pilots at Edwards during his time there, but few shared his insight into the physical capabilities of the projects on which he worked. This insight was put to the test in 1956, early in his NACA career, as co-pilot of a four-prop Boeing B-29 Superfortress. When one engine exploded at high altitude and sent fragments into two others, Armstrong and his commander, Stan Butchart, had to bring the huge aircraft safely to Earth on just one engine.

Throughout his time at Edwards, space was calling. When the USSR launched the world's first satellite, *Sputnik 1*, in 1957 and the world tracked its beeping orbit, it badly dented American pride. The US became determined to be the first to get a human being into space, and in 1958 Armstrong joined Man In Space Soonest (MISS), the USAF's new program for manned space flight.

Later the same year the National Aeronautics and Space Administration was created, swallowing up NACA and several research and test facilities. In 1962 Armstrong was picked to be a pilot for a new military space plane, the X-20 Dyna-Soar, for which he had

been an engineering consultant since 1960. Although the Dyna-Soar project was canceled in favor of the Gemini and Apollo programs, it laid the groundwork for the future success of the Space Shuttle.

Armstrong transferred to NASA's Astronaut Corps toward the end of 1962. He was by now a prominent figure in America's space program, one of the select band of men picked to lead their country's extraterrestrial adventure; and it was widely expected that he would be the US's first civilian astronaut. His chance came when he was named as command pilot for the *Gemini 8* mission.

It was to be the most ambitious mission of America's space program yet, the first ever attempt to

■ **ABOVE: Dave Scott and Neil Armstrong sit in *Gemini 8* atop their booster at Cape Kennedy, March 16, 1966.**

rendezvous and dock with a second, unmanned vehicle, called *Agena*. The flight was also to include a spacewalk by Armstrong's co-pilot David Scott. On March 16, 1966 the launch, rendezvous, and docking all went smoothly. But during a period of radio blackout (the result of gaps between tracking stations back on Earth) the two vehicles, now locked together, began to roll uncontrollably. Out of contact with ground control, Armstrong attempted a series of procedures to halt the movement which, at its worst, was spinning his capsule once every second.

When even undocking from the *Agena* failed to correct the problem,

it was clear that there was a failure in *Gemini*'s own maneuvering system. The only remaining option was to switch to the capsule's re-entry control – and that meant aborting the rest of the mission, including Scott's extra-vehicular activity. Safely back on Earth, Armstrong was criticized for not following malfunction procedures in his decision to abort but flight director Gene Kranz defended him. "The crew reacted as they were trained," he insisted. "And they reacted wrong because we trained them wrong."

Armstrong served as back-up commander on *Gemini 11* and *Apollo 8*, at the same time fulfilling a busy

program as an ambassador for the US space program both at home and abroad. In 1966 he accompanied President Johnson on a South American goodwill tour and in 1967 (on the day of the *Apollo 1* disaster) he was present in Washington DC at the launch of an international treaty

■ ABOVE: A portrait of the prime crew of the *Apollo 11* mission. From left to right they are: commander Neil A. Armstrong, command-module pilot Michael Collins, and lunar-module pilot Edwin E. Aldrin, Jr.

outlining legal protocols in space, the so-called Outer Space Treaty.

Two days before Christmas 1968, Neil Armstrong learned that he was to be the commander of *Apollo 11*, alongside lunar-module pilot Edwin "Buzz" Aldrin, and command-module pilot Michael Collins. The mission was simple – to land an American astronaut on the Moon – and the crew was the first in which all its

members had previous space-flight experience.

Once again, Armstrong cheated death during his training for the flight. During a simulation of the lunar landing in the so-called Flying Bedstead, he was hovering 100 feet up in the air when the controls failed. He would have crashed to Earth had he not judged to perfection the moment at which to eject. Once

the twentieth century, particularly in the light of the technical limitations of the age. Professor André Balogh of London's Imperial College, speaking 40 years later, reflected that "It was carried out in a technically brilliant way with risks taken… that would be inconceivable in the risk-averse world of today."

When, during the descent, he noticed that the planned landing site was covered in large boulders, Armstrong took over the controls of the lunar module in search of smoother terrain. The adjustment used more fuel than planned, a cause for concern at mission control. When Armstrong was able at last to report, "Houston, Tranquility Base here – the Eagle has landed," Houston replied, "You got a bunch of guys about to turn blue! We're

■ **BELOW:** Neil Armstrong and Buzz Aldrin work with special tools as they practice lunar surface activities. Armstrong is using a scoop device to pick up soil and rock samples from a simulated moonscape for storage in a sack held by Aldrin.

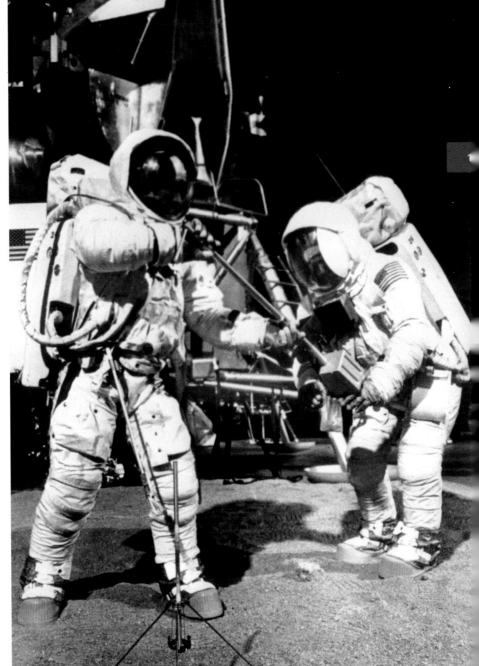

more, Armstrong's understanding of the physical behavior of his vehicle had saved his life.

Apollo 11 was launched from the Kennedy Space Center in Florida on July 16, 1969. Four days and a little under seven hours later, the lunar module, callsign Eagle, touched down on the surface of the Moon. It was without doubt the greatest scientific achievement of mankind in

Men on the Moon

breathing again."

After landing, the mission schedule called for the astronauts to sleep for five hours at the end of a long day. Understandably, however, no one was in the mood to rest and Armstrong asked that the next stage of the mission be brought forward. Preparations for it took six hours; they then depressurized the Eagle cabin and opened the hatch.

NASA had made the decision six months earlier that Armstrong should be the first to emerge – not only because he was the commander of the mission, but because, as a naturally quiet man, he would not let the honor go to his head. Now he climbed down the ladder, and paused. "I'm going to step off the LEM [lunar module] now," he reported. Then he turned and, at 2.56am UTC on July 21, 1969, set foot on the Moon's surface.

"That's one small step for man, one giant leap for mankind." The words were not scripted; they had come to him during the six hours of preparation after landing. It is estimated that a global television audience of some 600 million heard him say them in the live broadcast of the moment that changed mankind's relationship with its home planet forever.

Twenty minutes later Buzz Aldrin joined Neil. Together they planted a flag and set out a number of scientific instruments. There was time to take a short phone call from President Nixon and, after two and a half hours, Neil's last act was to leave a bag of items in memory of five space pioneers: Gus Grissom, Ed White, and Roger Chaffee who died in *Apollo 1*, and two Soviet cosmonauts – Vladimir Komarov and the first man in space, Yuri Gagarin. It was a thoughtful mark of respect from the modest first man on the Moon.

The three *Apollo 11* astronauts

■ OPPOSITE: *Apollo 11* astronauts Neil Armstrong, Michael Collins, and Buzz Aldrin are riding in this spacecraft as it lifts off the pad at Cape Kennedy.

received ticker-tape parades in New York, Chicago, and Los Angeles. Soon after his return, Armstrong declared that he would not go to space again. He had taken the first steps there – how could he follow that? He resigned from NASA in 1971, and spent the next eight years as Professor of Aeronautics Engineering at the University of Cincinnati.

Despite constant requests from politicians eager to bathe in his reflected glory, Armstrong (who married twice and had three children) never involved himself in politics. A spiritual but not a religious man, he shunned celebrity and stopped signing autographs when he discovered that forgeries were changing hands for high prices.

He died on August 25, 2012, from complications after heart surgery. Michael Collins, in tribute, said, "He was the best, and I will miss him terribly." Buzz Aldrin described him as "a true American hero and the best pilot I ever knew." The White House called his moonwalk "a moment of human achievement that will never be forgotten." He did an amazing thing, and he did it first.

■ **BELOW:** *Apollo 11* astronauts wave as they are carried through a deluge of tickertape and confetti in lower Manhattan, New York.

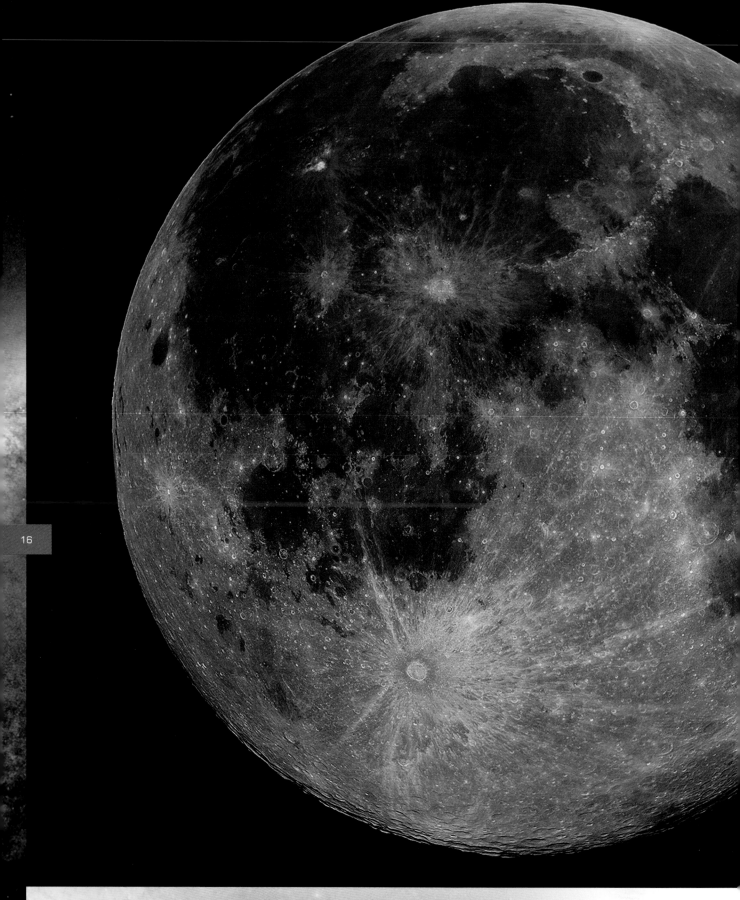

16

The Race to the Moon

"One small step for man, one giant leap for mankind," said Neil Armstrong, as he became the first human to walk on the surface of the Moon. But perhaps mankind's biggest leap was one of imagination.

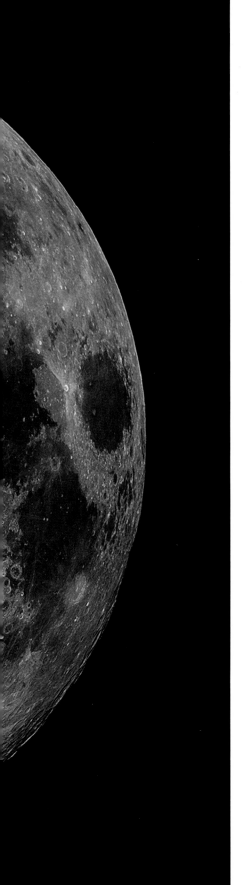

■ ABOVE: German rocket expert Wernher von Braun, 1955.

Our journey to the Moon began with the thought that it was not just a nocturnal light in the sky but a real place, and a possible destination.

It took mankind nearly 350 years to turn our fascination with the idea into triumphant reality. In 1625 a German scientist called Johannes Kepler published *The Dream*, a fantasy about being transported to the Moon and looking back to Earth. Konstantin Tsiolkovsky, a Russian school teacher, became intrigued after reading Jules Verne's 1865 sci-fi tale *From The Earth To The Moon*. Tsiolkovsky's theories of rocket transport anticipated many of the practicalities of space travel including

■ ABOVE: Laika, the female dog that was a passenger aboard *Sputnik 2*.

■ ABOVE RIGHT: Yuri Gagarin getting ready to become the first man to orbit the Earth.

multi-stage vehicles, liquid oxygen-hydrogen fuel, and even a winged ship uncannily like the Space Shuttle.

By the early twentieth century several individual pioneers were experimenting with rocket propulsion, both in theory and in practice. In 1926 Massachusetts physicist Robert Goddard, who had also been inspired by reading science fiction, launched the first liquid-fuel rocket to a height of 41 feet. A German school teacher, Hermann Oberth, had published a study paper in 1923 about human space travel; by 1930 the Society for Spaceship Travel in Berlin, founded by the technical advisers on a sci-fi movie called *The Woman In The Moon*, was also launching test rockets.

One of the Berlin society's members was a young enthusiast called Wernher von Braun who accepted the German Army's offer of work in 1933. Working in a top-secret facility, with 6,000 Nazi scientists and a workforce of concentration-camp prisoners, he developed the V-2 flying bomb in 1944. His genius for rocket technology made him a very desirable prize when World War II

ended, and when he surrendered to American troops in 1945, he soon found himself put to work in the development of US ballistic missiles.

Moscow's own rocket society, the Group for the Study of Reaction Motion, began practical launches in 1931. Founder member Sergei Korolëv, sent to a labor camp on a fabricated charge, was put to work on the USSR's rocket program in 1944. While Von Braun was working on America's Redstone and Jupiter rocket program, Korolëv developed the R-7 rocket which, in 1957, flew 3,700 miles, the world's first intercontinental ballistic missile.

1957-58 was designated International Geophysical Year, a celebration of scientific developments around the world, and both the US and the USSR proposed to launch a satellite during the 18-month festival. America's efforts were divided between the competing teams of its army and navy; both were caught by surprise when Russia's *Sputnik 1* satellite, the world's first, was sent into orbit aboard an adapted R-7. Against the backdrop of the Cold War, this was a very public

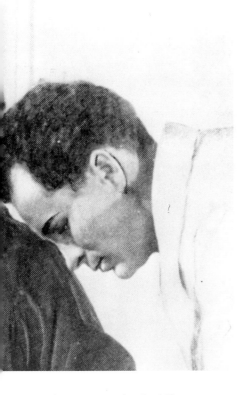

embarrassment for the US.

The situation was made worse when *Sputnik 2* was launched a month later carrying Laika the dog, the first living creature in space. Worse still, a month after that the US Navy's Vanguard launcher exploded on lift-off; and the navy's embarrassment was deepened when at last, in January 1958, the army's *Explorer 1* became the first US satellite in orbit. In March, the navy finally got *Vanguard 1* into space, but their success was soon eclipsed by *Sputnik 3*, which took off in May. The race was on, and the USSR was winning.

The next hurdle in the race was to put a man in space, and President Eisenhower created NASA to oversee America's effort. A team of potential astronauts was selected, known as the Mercury Seven after the capsule expected to carry them beyond the atmosphere. In September 1959, while NASA was still testing X-15 rocket planes, the Soviet Union dropped *Luna 2*, an unmanned craft, onto the lunar surface, and two weeks later *Luna 3* took the first ever photographs of the dark side of the Moon.

There was some comfort for the US when Ham the chimpanzee survived a 16-minute Mercury space flight in January 1961. But, on April 12, the Soviet spaceship *Vostok 1* took off from Kazakhstan carrying cosmonaut Yuri Gagarin, the first man in space, on a 108-minute flight around the Earth. America's Alan Shepard was very much in Gagarin's shadow when *Freedom 7* carried him into space for 15 minutes on May 5.

America was being comprehensively beaten at every step of the space race. The only way back was to dramatically raise the stakes, to give NASA time to catch up and overtake the Soviet Union. Three weeks after Alan Shepard's flight, on May 25, 1961, President Kennedy committed the US to landing a man on the Moon before the decade was out.

In August 1961, *Vostok 2* became the first spacecraft to orbit for over 24 hours. A full seven months later,

■ ABOVE: **Ed White moves away from his *Gemini 4* capsule as his golden tether unreels from a black bag in which it was kept until he emerged from the spacecraft. He somersaulted away for most of the tether's length, then maneuvered back with the aid of an oxygen rocket gun. This picture was taken by White's fellow astronaut, James McDivitt, June 8, 1965.**

John Glenn completed the first American orbits of the Earth, three of them in five hours in February 1962. But the Soviet Union continued to steal the limelight with longer flights and, in June 1963, the first female space traveler, cosmonaut Valentina Tereshkova.

Over 1962 and 1963 two further groups of US astronauts were selected to crew a new schedule of space flights under the Gemini and Apollo programs. Russia was still leading the race – in October 1964 *Voskhod 1* put the first three-man crew into space, and Alexei Leonov made the first ever spacewalk from *Voskhod 2* on March 18, 1965. But only a week later the first US manned flight, *Gemini 3*, took off; and on June 3, *Gemini 4* astronaut Ed White became the first US man to float in space.

■ ABOVE: *Apollo 1* astronauts, from left, Virgil "Gus" Grissom, Edward White II, and Roger Chaffee pose next to their Saturn 1 launch vehicle at Cape Canaveral Air Station prior to the launch – disaster was about to happen.

■ OPPOSITE: Moon photo taken by *Apollo 8*, December 1968.

America was catching up. In March 1966, it achieved the first ever docking of two spacecraft during the *Gemini 8* mission, an important procedure for any eventual Moon landing. Only four months after the USSR's *Luna 9* made the first controlled landing on the Moon, the first of seven unmanned Surveyor probes touched down there for the US in June 1966. In August that year, the first of five US Lunar Orbiters began to map the Moon's surface in preparation for a manned landing.

The Apollo program began in 1967 with a disastrous setback – the death of all three astronauts on board *Apollo 1* during a launch simulation on January 27. It was a rare failure in NASA's well-publicized and carefully-staged progress toward the Moon. The Soviet Union, by contrast, conducted its space

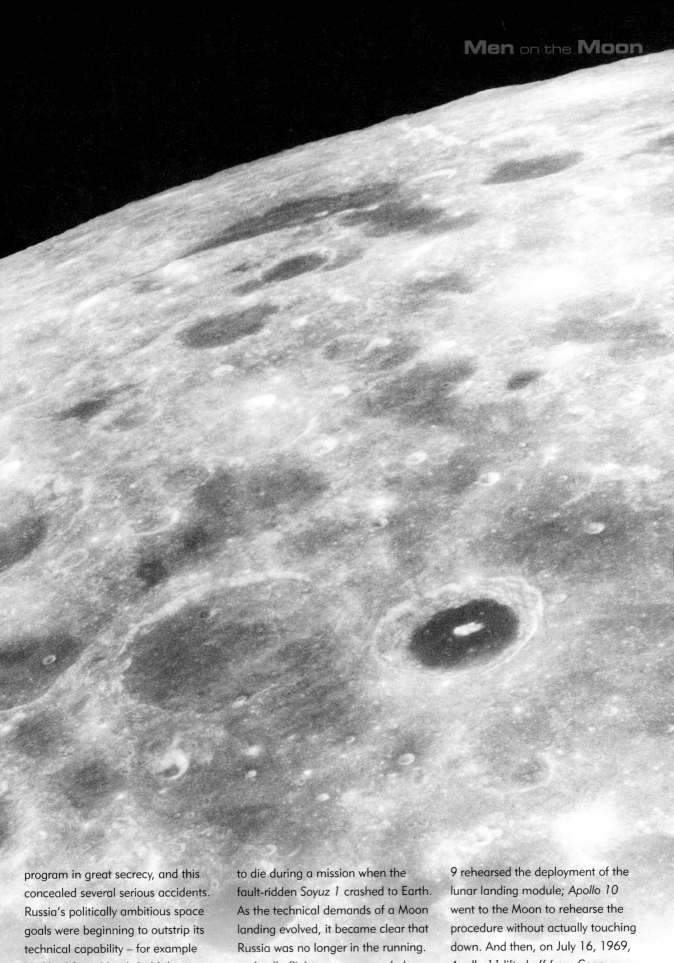

program in great secrecy, and this concealed several serious accidents. Russia's politically ambitious space goals were beginning to outstrip its technical capability – for example *Voskhod 1* could only hold three men by going without space suits and ejector seats; and three months after the *Apollo 1* deaths, Vladimir Komarov became the first spaceman to die during a mission when the fault-ridden *Soyuz 1* crashed to Earth. As the technical demands of a Moon landing evolved, it became clear that Russia was no longer in the running.

Apollo flights were suspended until the end of 1967, when *Apollo 8* made the first manned flight to the Moon, which it orbited 10 times. In Earth orbit during 1968, *Apollo 9* rehearsed the deployment of the lunar landing module; *Apollo 10* went to the Moon to rehearse the procedure without actually touching down. And then, on July 16, 1969, *Apollo 11* lifted off from Cape Kennedy. Four days later the lunar module touched down in the Sea of Tranquility; and in the small hours of July 21, 1969 (UTC), watched back

■ ABOVE: *Apollo 8* module pilot Jim Lovell, left, points to a lunar relief picture at Cape Kennedy as preparations for a blast off for a Moon orbit continued. Behind Lovell is William Anders (checkered shirt) and commander Frank Borman (right background) the other members of the trio scheduled to make the flight. At right is John Healey, a space project engineer, with Harold Collins, of Moon, NASA, at left background.

■ RIGHT: At NASA Mission Control in Houston's Manned Spacecraft Center, Donald "Deke" Slayton, left, director of flight crew operations, holds lithium hydroxide canisters attached to a hose as he discusses a makeshift repair to reduce the dangerous levels of carbon dioxide aboard the crippled spacecraft *Apollo 13*, April 15, 1970.

on Earth by 600 million viewers, Neil Armstrong took his one small step.

It was mankind's greatest achievement, the product of countless experiments, heroic failures, and inspired successes, by men and women all over the world dedicated to making the dream of traveling to the Moon a reality. The school teacher Konstantin Tsiolkovsky had summed it up neatly 66 years earlier: "First comes inevitably the idea, the fantasy, the fairy tale. Then comes

scientific calculation. Ultimately, fulfillment crowns the dream."

Six more Apollo missions set off for the surface of the Moon; only *Apollo 13* was forced to return without landing, when an exploding oxygen tank on April 13, 1970 endangered the lives of its crew. Lunar rover vehicles used from 1971 allowed astronauts to explore even more of the Moon's geography. In all, 12 astronauts, all American, walked on lunar soil, the last of them returning to Earth aboard *Apollo 17* on

December 14, 1972 after 22 hours on the surface. Russia meanwhile focused its efforts on establishing a space station, gaining valuable experience for future international efforts in that sphere of activity.

There was no doubt about the American public's early support for NASA's Moon program, particularly in the context of competition with its Cold War enemy the Soviet Union. But once the US won the space race, interest began to fade. Although the drama of the *Apollo*

■ ABOVE: The Earth-orbiting space shuttle *Challenger* beyond the Earth's horizon.

13 mission briefly revived interest, Congress's budgetary priorities were changing. In the end, it was financial constraints, rather than technical limitations, which made further journeys to the Moon impossible. Instead, NASA concentrated on unmanned exploration of the other planets in the Solar System and a reusable sub-orbital spacecraft, the Space Shuttle.

In 2004, President George W. Bush announced a plan to return to the Moon by 2020, with the aim of establishing a permanent base from which to launch manned missions further into space. In the wake of the economic collapse that overtook the world four years later, the timetable has become uncertain. But there can be no doubt that mankind will return to the Moon, pursuing the dream which has drawn him there since he began to imagine the possibility of such a journey.

Is it a dream worth pursuing? Take it from one who knows: Neil Armstrong, in an interview in 2005, said, "It's an interesting place to be. I recommend it."

Timeline of Space Exploration

c1625
German scientist Johannes Kepler writes the first work of science fiction, *Somnium* [*The Dream*], about a journey to the Moon

1865
French author Jules Verne publishes *From The Earth To The Moon*, which inspired modern scientists to take the possibility of Moon travel seriously

1903
Russian school teacher Konstantin Tsiolkovsky publishes *The Exploration Of Space With Reaction-Propelled Devices*, having researched the theory of oxygen-hydrogen rockets since the 1880s

1912
French aviator Robert Esnault-Pelterie gives talks on the possibility of space travel

1919
US physicist Robert Goddard declares that a solid-fuel rocket can reach the Moon

1923
German teacher Hermann Oberth's paper *Die Rakete Zu Den Planetenräumen* [*Rocket Into Interplanetary Space*] discusses human space flight with liquid propellants

March 16, 1926
In Auburn, Massachusetts, Robert Goddard launches the first liquid-fuel rocket – it flies 184 feet at a height of 41 feet

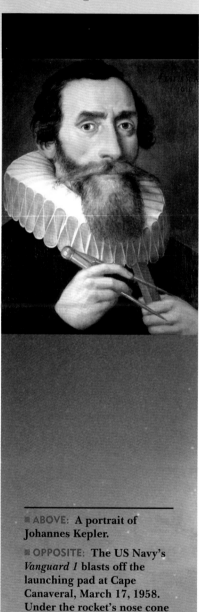

■ ABOVE: **A portrait of Johannes Kepler.**

■ OPPOSITE: **The US Navy's *Vanguard 1* blasts off the launching pad at Cape Canaveral, March 17, 1958. Under the rocket's nose cone is a 6.2-inch test satellite to be released into orbit around the Earth.**

January 25, 1930
In Berlin, the Verein für Raumschifffahrt [Society for Spaceship Travel] launches its first liquid-fuel rocket on a five-minute flight. Among its members is Wernher von Braun

1932
MosGIRD [the Moscow Group for the Study of Reaction Motion], one of several Soviet rocket clubs, sends a rocket to a height of 1,300 feet – one of its members is Sergei Korolëv

1934
The American Rocket Society, formed by members of science-fiction group the American Interplanetary Society, devotes itself to more practical experiments

September 1944
Start of Germany's bombing campaign using the V-2 rocket, the first man-made object to enter outer space, developed by Wernher von Braun

May 2, 1945
Von Braun surrenders to the US 44th Infantry Division as they advance through Germany. Issued with falsified papers, he is transferred to America a month later to work on the first US ballistic missile, Redstone

May 13, 1946
At Podlipki near Moscow, Stalin creates NII-88, Russia's space research facility, with Sergei Korolëv as the chief designer of long-range missiles

August 21, 1957
First long-distance flight, over
3,700 miles, of Sergei Korolëv's
R-7 Semyorka, the world's
first intercontinental ballistic
missile (ICBM)

October 4, 1957
Sputnik 1, first Soviet satellite, is
launched on a modified R-7 ICBM

November 3, 1957
First living creature in space, the
Samoyed terrier Laika ["Barker"], is
launched in *Sputnik 2*

January 31, 1958
Explorer 1, US Army's space project,
becomes the first US satellite

March 17, 1958
Vanguard 1, US Navy's space
project, is the second US satellite

May 15, 1958
Sputnik 3 is launched

October 1, 1958
NASA is founded to coordinate
America's space exploration program

April 9, 1959
The Mercury Seven – the first seven
US astronauts – are selected, among
them pilots of all US space projects,
from Mercury to the Space Shuttle.
They included John Glenn and
Alan Shepard

September 14, 1959
The Soviet Union's unmanned *Luna
2* spacecraft successfully lands on
the Moon

September 17, 1959
First supersonic flight of the North
American X-15 rocket plane, piloted
by Scott Crossfield

October 4, 1959

Russia's *Luna 3* takes the first ever photographs of the dark side of the Moon

December 4, 1959

Sam, the rhesus macaque, on board a Jupiter rocket, is the first of four animal test pilots in the Mercury program, followed over the next two years by Miss Sam, and chimpanzees Ham and Enos

April 12, 1961

Soviet cosmonaut Yuri Gagarin, with a 108-minute flight on board *Vostok 1*, is the first man in space

May 5, 1961

First US man in space, in *Freedom 7* (the first mission of the Mercury project), is Alan Shepard, carried by a Redstone rocket in a 15-minute flight

May 25, 1961

President Kennedy commits America to a Moon landing before the end of the decade

October 27, 1961

First launch of Wernher von Braun's Saturn 1 rocket, developments of which will eventually carry a man to the Moon

February 20, 1962

John Glenn lifts off on board *Friendship 7*, powered by an Atlas rocket, to become the first US astronaut to orbit the Earth – three times in less than five hours

■ **ABOVE:** Alan Shepard, the first American to journey into space, peers into his *Freedom 7* space capsule after it is recovered from the Atlantic Ocean.

■ **BELOW:** Rescue personnel work over the X-15 as it lies in the desert, minutes after an emergency landing, November 1959.

August 27, 1962

Launch of NASA's *Mariner 2*, first probe to visit another planet: Venus

September 17, 1962

The second group of US astronauts is selected, to man the newly announced Gemini and Apollo space programs – the nine include Ed White and Neil Armstrong

June 16, 1963

First woman in space is Soviet cosmonaut and former cotton mill worker, Valentina Tereshkova, in *Vostok 6*

October 1963

A third group of US astronauts is announced – among the 14 named are Buzz Aldrin and Michael Collins

March 18, 1965
First spacewalk, by Alexei Leonov from *Voskhod 2*

March 23, 1965
First of 10 manned flights in the Gemini program is *Gemini 3*, which carries two astronauts into space on a Titan II rocket

June 3, 1965
Gemini 4 astronaut Ed White is the first American to walk in space

February 3, 1966
Unmanned Soviet craft *Luna 9* makes the first controlled landing on the Moon

March 16, 1966
First ever docking of two spacecraft is achieved by Neil Armstrong and Dave Scott in *Gemini 8*

June 2, 1966
Unmanned US spacecraft *Surveyor 1* lands on the Moon, the first of seven in the Surveyor program to do so

August 10, 1966
Lunar Orbiter 1 lifts off, the first of five US probes which will eventually map the Moon's surface in the search for Apollo landing sites

January 27, 1967
All three astronauts in *Apollo 1* – Gus Grissom, Ed White, and Roger Chaffee – die when fire breaks out during a launch simulation

April 24, 1967
Vladimir Komarov is the first person to die during a spaceflight when Russia's hastily launched *Soyuz 1* crashes on its return to Earth

December 21, 1967
Apollo 8, the first manned flight to the Moon and the first to use a Saturn 5 rocket, takes off and makes 10 lunar orbits

July 20, 1969
First Moon landing by Neil Armstrong and Buzz Aldrin of *Apollo 11*, watched by Mike Collins in the command module orbiting the Moon

April 13, 1970
An oxygen tank explodes on board *Apollo 13*, forcing the cancellation of their Moon mission and a dangerous return journey to Earth by Jim Lovell,

Jack Swigert, and Fred Haise

April 19, 1971
The Soviet Union launches the first space station, *Salyut 1*

June 30, 1971
After three weeks in *Salyut 1*, the three-man crew of *Soyuz 11* are the first people to die in space when the air from their descent module leaks out

---

■ **ABOVE:** 26-year-old Valentina Tereshkova, who became the first woman to travel in space, is pictured as seen in a television transmission from her spacecraft, *Vostok 6*.

■ **BELOW:** This is an artist's concept of *Soyuz 11* and the *Salyut 1* spacecraft, preparing to dock in space, 1971.

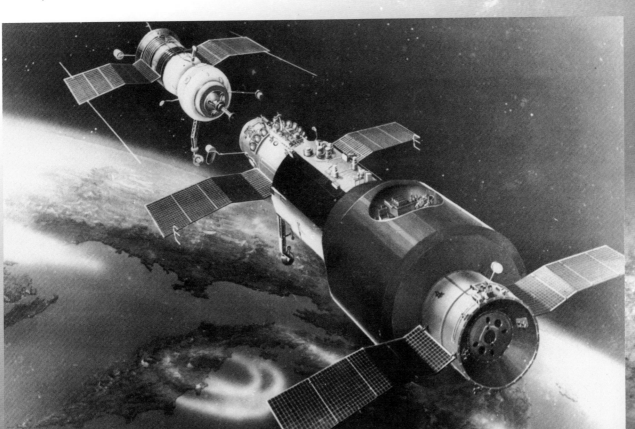

July 26, 1971

Apollo 15 blasts off carrying the first Lunar Roving Vehicle to the Moon

December 14, 1972

Eugene Cernan and Harrison Schmitt of *Apollo 17*, the last men to walk on the Moon for 40 years, take off from the lunar surface

May 14, 1973

NASA launches *Skylab*, an experimental orbiting station

February 8, 1974

The crew leaves *Skylab* for the last time

July 17, 1975

The symbolic rendezvous and docking of Russia's *Soyuz 19* and an American Apollo spacecraft marks the last flight by an Apollo vehicle

June 19, 1976

NASA's *Viking 1* deep space probe enters Mars' orbit 10 months after leaving Earth's – a month after surveying the planet, it will touch down on its surface, to be followed on September 3 by *Viking 2*

August 20, 1977

The *Voyager 2* space probe is launched to explore the outer limits of the Solar System – Jupiter, Saturn, Uranus, and Neptune – before exploring interstellar space

■ **ABOVE:** *Apollo 15* **commander Dave Scott, command-module pilot Alfred Worden, and lunar-module pilot James Irwin, enroute to their spacecraft for a countdown demonstration test in Cape Kennedy, Florida, 1971.**

■ **RIGHT:** **The first color picture taken on the surface of Mars, 1976, by the** *Viking 1* **lander, shows that the Martian soil consists mainly of reddish fine-grained material. However, small patches of black or blue-black soil were found deposited around many of the foreground rocks.**

■ **ABOVE: Space Shuttle *Columbia* is lifted off the launch pad at Cape Canaveral, making the first flight of this reusable spacecraft, at Kennedy Space Center, 1981.**

September 5, 1977
NASA's *Voyager 1* sets off on a faster trajectory, visiting only Jupiter and Saturn; by 2012 it is on the very edge of the Solar System

December 24, 1979
First launch of the European Space Agency's *Ariane 1* rocket ushers in a new era of relatively cheap satellite delivery

April 12, 1981
First flight of the Space Shuttle *Columbia*

January 28, 1986
Space Shuttle *Challenger* explodes less than two minutes into its flight, killing all seven crewmembers

February 19, 1986
The base block of the third-generation Soviet space station *Mir* is launched from Kazakhstan, to be followed by other elements assembled in orbit over the next 10 years

May 4, 1989
Magellan, an unmanned probe that

will map Venus, is launched from the Space Shuttle *Atlantis*

April 24, 1990
The Hubble space telescope is deployed from Space Shuttle *Discovery* but a flawed mirror renders it almost useless

December 2, 1990
Soyuz TM-11 lifts off for the *Mir* space station in the first ever commercial space flight, carrying Toyohiro Akiyama, a Japanese TV reporter

December 12, 1993
Space Shuttle *Endeavour* begins its mission to repair the Hubble telescope

September 29, 1995
After 22 years of space exploration, *Pioneer 11* (originally launched to study Jupiter and Saturn) closes down, but continues its journey toward the constellation Scutum

December 7, 1995
The *Galileo* probe completes its six-year journey from the Space Shuttle *Atlantis* to Jupiter, where it will spend the next eight years surveying the planet

October 29, 1998
John Glenn returns to space on the Space Shuttle *Discovery* at the age of 77

November 20, 1998
Construction begins of the International Space Station, a cooperative effort by 16 countries

April 28, 2001
The first space tourist to pay for his own ticket, Dennis Tito of Santa Monica, California, takes off for the International Space Station on board a Soyuz-TM rocket – at a cost to him of $20 million

■ ABOVE: **The Hubble space telescope is seen in 2002 as the Space Shuttle *Columbia*, with a crew of seven astronauts on board, approached to latch its robotic arm onto the giant telescope.**

■ BELOW: Mars rover *Curiosity*.

January 23, 2003
Communication is finally lost with
Pioneer 10, launched on March 2,
1972 to study Jupiter, on its way to
the star Aldebaran

February 1, 2003
Space Shuttle *Columbia* disintegrates
during re-entry after its 28th mission,
killing all seven crewmembers

October 15, 2003
China becomes the third nation to
put a man in space when Yang Liwei
takes off on board *Shenzhou 5* to
orbit the Earth 14 times

December 25, 2003
British project *Beagle 2*, designed to
seek signs of life on Mars, disappears
without trace after deploying from its
mothership *Mars Express*

October 4, 2004
SpaceShipOne, designed by Burt
Rutan, wins the $10 million Ansari X
Prize as the first privately-built ship to
reach space twice within two weeks

January 19, 2006
NASA's *New Horizons* probe lifts off
on a journey passing Mars, Jupiter,
Saturn, and Uranus

October 24, 2007
China begins its Moon exploration
with the launch of *Chang'e 1*, an
unmanned lunar orbiter

April 28, 2008
India's Polar Satellite Launch Vehicle
C9 successfully deploys a record 10
satellites in one trip

August 6, 2012
Mars rover *Curiosity* lands on the
planet, where a month later it finds
evidence of water channels

2013
China plans to land unmanned rover
vehicles on the Moon

■ ABOVE: Senator John
Glenn, the world's oldest
astronaut, lands back on
Earth, 1998.

2013
India plans to launch *Mangalyaan*, a
Mars orbiter

2013
The first regular space tourism flights
are planned by two companies:
XCOR Aerospace, in a Lynx rocket
plane out of Midland, Texas, and
Virgin Galactic, using *SpaceShipTwo*

July 14, 2015
Estimated arrival of NASA's
New Horizons probe at its
destination, Pluto

2018
European Space Agency plans a
mission to Mars

2020
NASA plans to return to the Moon to
establish a permanent base

10 Top Space Men (and Women)

Michael Collins

Former test pilot Collins, born October 31, 1930, flew in space twice – firstly in *Gemini 10*, and more famously in 1969, as command-module pilot for *Apollo 11*, orbiting the Moon as Armstrong and Aldrin made the first manned landing. In his autobiography, he wrote that "this venture has been structured for three men, and I consider my third to be as necessary as either of the other two."

He had been inspired by John Glenn, and was accepted for astronaut training at the second attempt in 1963. Collins retired from NASA in the year following the lunar mission to work in the Department of State. In 1971, he became the director of the National Air and Space Museum and was appointed undersecretary of the Smithsonian Institution in 1978.

Yuri Gagarin

The first man in space, Gagarin (born March 9, 1934) was the son of a farmer, plucked from the Soviet air force to undergo cosmonaut training. He flew only one space mission, orbiting Earth in *Vostok 1* on April 12, 1961, at a speed of 17,025 mph for 108 minutes. At the highest point, he was about 200 miles above the Earth. The history-making Colonel Gagarin died on March 27, 1968, when the MiG jet fighter he was piloting crashed near Moscow. At the time of his death, he was in training for a second space mission.

■ ABOVE: Michael Collins.
■ BELOW: Yuri Gagarin.

■ ABOVE: **John Glenn.**
■ BELOW: **Gus Grissom.**

John Glenn

John Herschel Glenn, Jr., born July 18, 1921, in Cambridge, Ohio, was selected as a Project Mercury astronaut in 1959. He was the third American in space and first to orbit the Earth when he circled the globe three times in *Friendship 7* at more than 17,000 mph. He controlled nearly two of the orbits himself after problems with the automatic altitude control system. In 1998, he made history again when he became the oldest man ever to fly in space.

Glenn enjoyed a career in politics, ran for the Democratic vice-presidential nomination in 1976, and has publicly criticized the decision to end the Space Shuttle program.

Gus Grissom

Born on April 3, 1926, Virgil I. Grissom joined the United States Air Force in 1950 and served as a fighter pilot during the Korean War, earning the Distinguished Flying Cross. One of the original Mercury astronauts, he was the second American to fly in space, and the first member of the NASA Astronaut Corps to fly in space twice. Grissom died with Roger B. Chaffee and Edward White on January 27, 1967, at Cape Kennedy, Florida, during the testing of *Apollo 1* when fire broke out in the cockpit. He was buried in Arlington National Cemetery, beside Chaffee; sons Scott and Mark, teenagers at the time of the tragedy, followed in their father's footsteps to become pilots.

Vladimir Komarov

Test pilot and aerospace engineer Vladimir Komarov, born March 16, 1927, was the victim of an over-ambitious Soviet plan in 1967 to stage a mid-space rendezvous between two Soviet spaceships. Technicians who inspected *Soyuz 1* prior to launch found many structural problems and recommended postponement, but a memo to

Soviet leader Leonid Brezhnev from Komarov's friend and fellow astronaut, Yuri Gagarin, was never acted upon. Komarov was on board on April 24, 1967, and was killed on impact when the malfunctioning craft plummeted to Earth. It's said he knew his likely fate but would not back out as Gagarin (who ironically died the following year) would have gone in his place. His flight on *Soyuz 1* made him the first cosmonaut to fly into outer space more than once and the first human to die during a spaceflight.

Yang Liwei

China's first astronaut was born in Suizhong, Liaoning Province in June 1965. He joined the People's Liberation Army in 1983 and became a fighter pilot. In 1998, he joined the country's first team of astronauts.

Yang was launched into orbit aboard the spaceship *Shenzhou 5* on October 15, 2003. This orbited the Earth 14 times and traveled 310,685 miles before landing safely on the central grasslands of Inner Mongolia on the following morning. He is now deputy director of the China Astronaut Research and Training Center and also deputy director general of China's manned space program.

Jim Lovell

Former US Navy commander Lovell, born March 25, 1928, is the first of only three people to fly to the Moon twice, and the only one to have flown there twice without making a landing. Lovell was also first to fly in space four times, having accrued over 715 hours and seen 269 sunrises.

He led the aborted *Apollo 13* Moon mission in which he and fellow astronauts Jack Swigert and Fred Haise nearly died after an explosion on April 13, 1970, but a combination of ingenuity, endurance, and luck prevailed and the trio made it home

■ ABOVE: **Jim Lovell.**

■ OPPOSITE ABOVE: **Vladimir Komarov (center).**

■ OPPOSITE: **Yang Liwei.**

safely. A small crater on the far side of the Moon is named Lovell in his honor.

Christa McAuliffe

Christa McAuliffe, born September 2, 1948, was a civilian who died with the rest of the seven-member crew when *Challenger* exploded 73 seconds after launch on January 28, 1986. A junior high school social studies and history teacher from New Hampshire, she was selected from 11,500 applicants to teach two lessons from the space shuttle. When a guest on *The Tonight Show Starring Johnny Carson*, she stated, "If you're offered a seat on a rocket ship, don't ask what seat. Just get on."

In 1990, McAuliffe was portrayed by Karen Allen in the TV movie *Challenger*, while in 2004 she was posthumously awarded the Congressional Space Medal of Honor along with her fellow *Challenger* astronauts.

Alan Shepard

Born in 1923, Shepard became the first American in space on May 5, 1961, when he traveled 302 miles in the tiny Mercury space capsule *Freedom 7*, hitting a top speed of 5,100 mph before dropping 116 miles into the Atlantic. He was the world's second ever astronaut, only 23 days after Yuri Gagarin.

Shepard was taken off the Apollo Moon landing program with an inner ear infection but thereby missed the *Apollo 13* mission that ended in near-disaster. Ten years after his first flight, he returned to space for the second and last time as commander of *Apollo 14* and became not only the fifth person to walk on the surface of the Moon but also the then oldest at 47.

Valentina Tereshkova

Tereshkova was born to a peasant family on March 6, 1937 and combined work in a textile mill with amateur parachuting. She applied to become a cosmonaut in 1961 and, after 15 months' training, was chosen to fly aboard *Vostok 6*, launched on June 16, 1963. Her flight, which won her the Order of Lenin and Hero of the Soviet Union awards, lasted 48 orbits totaling 70 hours 50 minutes. The next woman flew in space in 1982.

She married fellow cosmonaut Andrian Nikolayev in November 1963; daughter Yelena was the first child of parents that had both been in space, but the couple later divorced. Tereshkova served as president of the Soviet Women's Committee and was a member of the Supreme Soviet, the USSR's national parliament.

■ ABOVE: Valentina Tereshkova.

■ ABOVE: Christa McAuliffe.
■ OPPOSITE: Alan Shepard.

Moonwalkers

The 12 people to have walked on the
Moon between 1969-72:

Neil Armstrong
Apollo 11, July 21, 1969

Edwin "Buzz" Aldrin
Apollo 11, July 21, 1969

Charles "Pete" Conrad
Apollo 12, November 19, 1969

Alan Bean
Apollo 12, November 19, 1969

Alan Shepard
Apollo 14, February 5, 1971

Edgar Mitchell
Apollo 14, February 5, 1971

David Scott
Apollo 15, July 31, 1971

James Irwin
Apollo 15, July 31, 1971

John W. Young
Apollo 16, April 21, 1972

Charles Duke
Apollo 16, April 21, 1972

Eugene Cernan
Apollo 17, December 11, 1972

Harrison Jack Schmitt
Apollo 17, December 11, 1972

Space Spin-offs

There are those who believe that the space race was a purely political race between two great superpowers: which country could go the furthest, the fastest, be the first. But the billions of dollars invested in space exploration have also had many beneficial scientific spin-offs, some of which we take for granted in our daily lives. We come into contact with NASA inventions when we walk home, visit the doctor and hospital, drive our cars, and enjoy our leisure time.

Medical Inventions

MRI/CAT Scanning

The lives of many people suffering from cancer and related diseases have been transformed by advances in medical scanning equipment. MRI (Magnetic Resonance Imaging) and CAT (Computer Axial Tomography) are derived from digital signal processing which was developed

■ **RIGHT: An MRI scanner being prepared for use.**

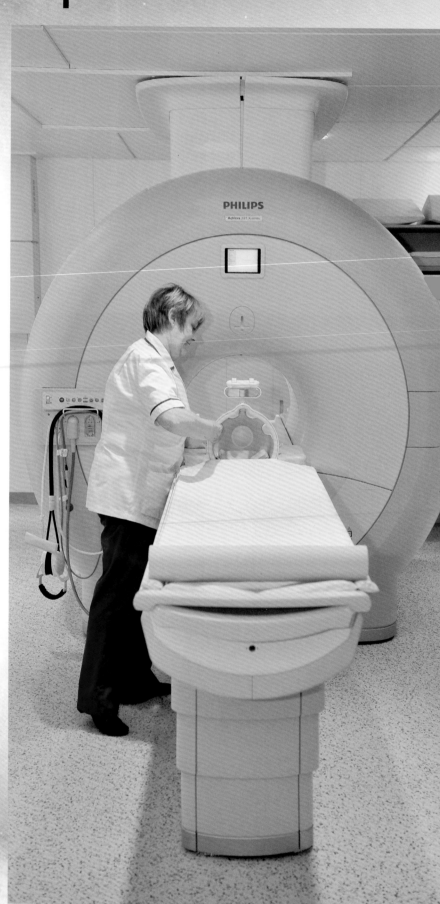

from the Apollo missions. NASA needed to digitally enhance pictures of the Moon, and also to detect tiny imperfections in rocket structures and components. Applying these technologies to medical scanning has resulted in much-improved diagnosis of soft-tissue diseases such as cancer.

Kidney Dialysis

One of the perennial problems on a long-distance space mission is – what do you do with waste products? NASA needed to develop technology to purify and recycle water. From this research came about a huge improvement in the lives of patients who require kidney dialysis – people who needed to use a machine to remove waste from their bodies because their kidneys are malfunctioning. The new machines require much less water to operate, and mean that patients have much more freedom of movement during treatment.

Laser Angioplasty

America's space program developed laser technology for the remote probing and analysis of the ozone layer. A company from Irvine, California took this technology and enhanced it to make a cool-running laser that uses ultraviolet light energy to perform very delicate and accurate surgery on human arteries. The laser is guided to the blockage, then used to destroy the blockage, layer by layer, vaporizing it into gaseous particles.

There are many other important medical advance spin-offs from the space program including cochlear implants which help profoundly deaf people to hear again, an artificial heart, blood analysis equipment, lighting technology which treats brain tumors in children, and more manageable insulin pumps for diabetics.

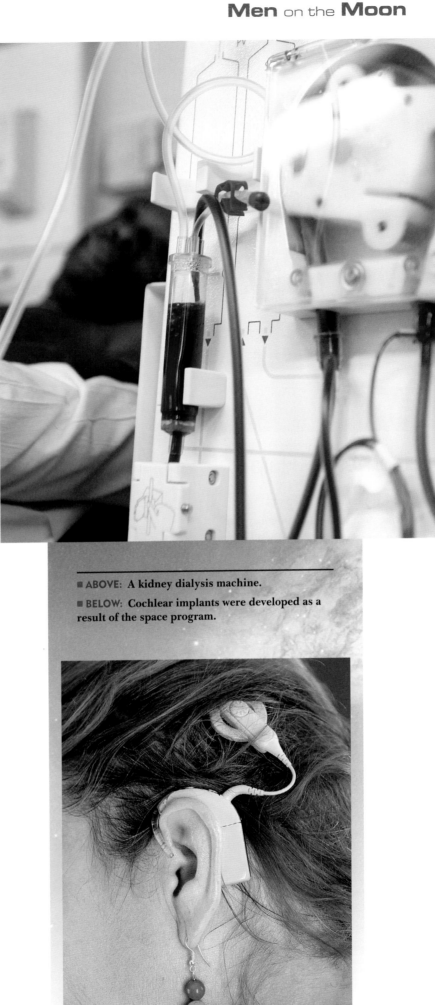

■ ABOVE: A kidney dialysis machine.

■ BELOW: Cochlear implants were developed as a result of the space program.

Domestic Help

Cordless Power Tools

Manufacturers had been trying to develop power tools that worked without a mains cable for some time, but were always held back by the weight and operating life of the battery needed. NASA was seeking lightweight and portable equipment for the astronauts to use when acquiring samples to bring home. Through this, they invented power tools that were small, easy to carry, and had an effective battery life. This has led to a massive change in the power-tool industry, and every home probably has a rechargeable screwdriver, drill, or even lawn mower they can use without the tyranny of a cord.

Memory Foam

NASA needed a material that would reduce the impact of take-offs and landings on the astronauts' bodies. In looking for a suitable "shock-absorbing" solution, they came up with memory foam. This foam molds to the shape of your body as it warms

■ ABOVE: **Cordless power tools charge without being plugged in, using space technology.**

■ BELOW: **A memory foam pillow is demonstrated.**

up and, in doing so, relieves pressure on bony areas like hips, knees, and neck.

Astronauts are under less strain as they travel to the stars, and the rest of us can enjoy a better night's sleep.

Scratch-resistant Lenses

NASA developed scratch-resistant lenses when they were trying to protect astronauts' visors and keep them fully transparent. They invented a process called "dual ion-beam

■ **LEFT:** A lot of spectacles now come as standard with scratch-resistant lenses.

■ **BELOW:** Hamakawa Elementary School children look at packs of space food during a special "space" class in Tokyo. Assisted by the Japan Aerospace Exploration Agency, the company, which has been developing with NASA a next-generation airless tire for use in lunar exploration, offered students an opportunity to study replicas of rockets, satellites, space suits, and space food products.

4

bonding" which involves applying a diamond-like lubricating coat over plastic lenses. Today, the coating is still used to prolong the life of the hard resin plastic in eyeglasses people wear every day.

Freeze-dried Food

All humans have to eat, and astronauts are no exception. To keep food fresh on the long voyage to the Moon NASA adopted the technique of freeze-drying, employing an earlier invention that had been used to preserve plasma for medical purposes in World War II. In the process, the food is quickly frozen and the water evaporated by turning it into vapor.

Food prepared in this way is sealed in vacuum packs and does not need to be kept frozen. It is reconstituted using simple hot water. Astronauts even took freeze-dried ice cream into space. This can be kept at room temperature without melting and is more brittle and rigid than standard ice cream but is still soft when bitten into. Freeze-dried food is in common use today in, for example, instant coffee and supplies carried by walkers and expeditionary groups.

Safety First

Firefighting Equipment

The safety of fire fighters has had more than one boost due to space research. Firstly, the science behind life-support systems for astronauts enabled the development of breathing apparatus much smaller and lighter than previously. It included a new facemask that gave better peripheral vision and an air-depletion warning. There have since been far fewer smoke-inhalation injuries to firemen.

More than this, after a fire on the Apollo launchpad in 1967 resulted in the deaths of three astronauts, a line of fire-resistant materials were developed for use in space suits and vehicles. Today these are used not only for greater protection when fire fighting, but also in military, motor sports, and many other applications.

Safety Grooving

NASA was worried about the problems of water on the runway causing the space shuttle to hydroplane on landing. In an attempt to reduce this dangerous effect, they found making narrow grooves in the tarmac forced water away from the surface. This simple engineering technique is now used on interstates, sidewalks, playgrounds, and station platforms and has been estimated to have reduced wet-weather road accidents by 85 per cent.

Shock-absorbing Soles

The low gravity on the Moon meant astronauts would take giant steps, possibly overbalancing or landing awkwardly. A shock-absorbing material was designed to assist stability and motion control, and these materials have since been incorporated into sneakers and running shoes to make them more comfortable for athletic use.

Cool Suits

Cooling suits, developed to keep Apollo astronauts comfortable during Moon walks, are today worn by racing drivers, people with multiple sclerosis, and children with a congenital disorder known as hypohidrotic ectodermal dysplasia which restricts the body's ability to cool itself.

■ BELOW: Firefighters are protected by suits developed by space technology.

Waves from Space!

Much of what we watch, listen to, and use to help us get around comes from satellites orbiting the Earth. But managing those signals effectively was a problem the space program helped improve.

Satellite navigation or GPS (Global Positioning System) came about through a military project on space-based navigation systems. Today it is used in military applications, but also in civil aviation, automobile, and sea transportation to stop people getting lost. It also forms the basis of many surveying, mapping, and meteorological projects.

NASA developed ways to correct errors in the signals coming from their spacecraft. This technology is used to reduce noise (compromised picture or sound) in TV signals coming from satellites and make viewing more rewarding.

Space Invention Trivia

Space Pens

Most pens depend on gravity to make the ink flow into the ballpoint. The Fisher Space Pen's ink cartridge contains pressured gas to push the ink toward the ballpoint. It also uses special ink that works in very hot and very cold environments.

Aerodynamic Bicycle Wheel

A special sporting bike wheel benefits from NASA research into shuttle wings. Three spokes on the wheel act like wings, making the bicycle very efficient for racing.

Failsafe Flashlight

This uses NASA's concept of system redundancy – always having a backup for the parts of the spacecraft with the most important jobs. The flashlight has an extra-bright primary bulb and an independent backup

■ ABOVE: Modern GPS systems were developed following a military project on space-based navigation systems.

system with its own separate lithium battery (also NASA-developed technology) and its own bulb.

Since the Fifties, when the space race began in earnest, NASA alone has pioneered more than 6,000 important scientific improvements, which have changed the way we live our lives. Couple this with inventions achieved through space-flight research in Russia and in Europe, and it is clear just how important the space program has been to our lives.

Our Solar System

NASA
solarsystem.nasa.gov

Exploring the Planets

While the Moon was the nearest, and most obvious, world beyond Earth on which to take our tentative first extra-terrestrial steps, the space-race countries already had their eyes on other planets. Exploring the planets helps to develop our understanding of how life on Earth came to be.

The first successful interplanetary flyby was the 1962 *Mariner 2* flyby of Venus. Flybys for the other planets were first achieved in 1965 for Mars by *Mariner 4*, 1973 for Jupiter by *Pioneer 10*, 1974 for Mercury by

■ ABOVE: **A montage of the planets and bodies in our Solar System.**

Mariner 10, 1979 for Saturn by *Pioneer 11*, 1986 for Uranus by *Voyager 2*, and 1989 for Neptune by *Voyager 2*.

The first interplanetary mission to send home surface data from another planet was the 1970 landing of *Venera 7* on Venus which returned 23 minutes of data to Earth. Over two hours of transmission was made from the surface of Venus by *Venera 13* in 1982, the longest ever Soviet planetary surface mission.

The closest planet to the Sun, Mercury is the least explored of the inner planets. *Mariner 10* and MESSENGER missions have been the only missions that have made close observations of Mercury. MESSENGER entered orbit around Mercury in March 2011, to further investigate the observations made by *Mariner 10* in 1974-75. A third mission to Mercury, *Mariner 10* and MESSENGER missions (a joint mission between Japan and the European Space Agency) is scheduled to arrive in 2020.

Venus has long fascinated scientists, partly because of its similarity in size and proximity to the Earth, and partly because it is so obscured by clouds. Could life be hidden under them? It was the first target of interplanetary flyby missions and has had more landers sent to it (nearly all from the Soviet Union) than any other planet in the Solar System. The first successful Venus flyby was the American *Mariner 2* spacecraft, which flew past Venus in 1962. In 1967, *Venera 4* became the first probe to enter and directly examine the atmosphere of Venus. In 1970, *Venera 7* was the first successful lander to reach the surface of Venus, and by 1985 it had been followed by eight additional successful Soviet Venus landers, which provided images and other direct surface data. Ten further successful orbiter missions have been

■ ABOVE: This is a *Mariner 2* spacecraft that was hurled aloft from Cape Canaveral, 1962, on a 109-day flight toward the planet Venus. An hour after the launch, a tracking station at Woomera, Australia, picked up signals that indicated the spacecraft had unfolded properly and that the solar panels were supplying power to the craft's batteries.

sent to Venus, starting in 1975 with the Soviet orbiter *Venera 9*.

The exploration of Mars has been an important part of the space programs of the Soviet Union (later Russia), the United States, Europe, and Japan. Many robotic spacecraft have been launched toward Mars since the Sixties. In 1971, the *Mars 3* mission achieved the first soft landing on Mars, returning data for almost 20 seconds. Later, much longer duration surface missions were achieved, including over six years of Mars surface operation by *Viking 1* from 1975 to 1982.

The investigation of Jupiter has consisted of a number of automated NASA spacecraft visiting the planet since 1973. Saturn has been explored only through unmanned spacecraft launched by NASA, including one mission (Cassini-Huygens) planned and executed in cooperation with other space agencies. These missions consist of flybys in 1979 by *Pioneer 11*, in 1980 by *Voyager 1*, in 1982 by *Voyager 2*, and an orbital mission by the *Cassini* spacecraft, which entered orbit in 2004.

The exploration of Uranus has been entirely through the *Voyager 2* spacecraft, with no other visits currently planned. The closest approach to Uranus occurred on January 24, 1986. *Voyager 2* studied the planet's unique atmosphere and magnetosphere. *Voyager 2* also examined its ring system and the moons of Uranus including all five of the previously known moons, while discovering an additional 10 previously unknown moons.

As of 2012, *Voyager 1* is the farthest man-made object that has ever been sent from the Earth. NASA has reported it may be very close to entering interstellar space and becoming the first man-made object to leave the Solar System.

■ **BELOW:** The unmanned *Voyager 1* sent back pictures of the planet Saturn to Earth in 1980. It has since traveled further than any man-made craft.

Mercury

Distance from the Sun	36 million miles.
Distance from Earth	57 million miles.
Size	Volume: 0.056 x Earth Mass: 0.055 x Earth.
Gravity	If you weigh 100 pounds on Earth, you would weigh 38 pounds on Mercury.
Length of year	88 Earth days.
Length of day	59 Earth days.
Surface temperature	Temperatures on Mercury's surface can reach 800°F. Because Mercury's atmosphere is so thin, the surface cannot retain that heat so nighttime temperatures can drop to -290°F.
Atmosphere	Mercury's thin atmosphere, or exosphere, is made up of atoms blasted off the surface by the solar wind and micrometeoroid impacts. 95 per cent helium and some hydrogen.
First photographed	**1974-1975:** *Mariner 10* photographed roughly half of Mercury's surface in three flybys.
First landings	Not landed on yet. **2008:** MESSENGERs carried out three flybys, which revealed the side of the planet not seen by *Mariner 10*.
Topology	Mercury's surface resembles that of Earth's Moon, scarred by many impact craters resulting from collisions with meteoroids and comets. While there are areas of smooth terrain, there are also lobe-shaped scarps or cliffs, some hundreds of miles long and soaring up to a mile high, formed by contraction of the crust.
Moons	None.
Interesting factoid	Mercury's magnetic field has just 1 percent the strength of Earth, but the field is very active.

48

■ ABOVE: A portion of western Eistla Regio is displayed in this three-dimensional perspective view of the surface of Venus. This is a computer-generated image based on radar data collected by the US *Magellan* spacecraft, combined with color images from the Soviet *Venera 13* and *14* spacecraft. The image was released by NASA's Jet Propulsion Laboratory, 1991.

Venus

Distance from the Sun	67.24 million miles.
Distance from Earth	At its closest, Venus is 26 million miles away.
Size	About 650 miles smaller in diameter than Earth Volume: 0.857 x Earth Mass: 0.815 x Earth.
Gravity	If you weigh 100 pounds on Earth, you would weigh 88 pounds on Venus.
Length of year	225 Earth days.
Length of day	243 Earth days.
Surface temperature	Ranges from 55°F to 396°F.
Atmosphere	Carbon dioxide (95 percent), nitrogen, sulfuric acid, and traces of other elements.
First photographed	1962: NASA's *Mariner 2* reached Venus and revealed the planet's extreme surface temperatures. It was the first spacecraft to send back information from another planet.
First landings	1970: The Soviet Union's *Venera 7* sent back 23 minutes of data from the surface of Venus. It was the first spacecraft to land successfully on another planet.
Topology	A rocky, dusty, waterless expanse of mountains, canyons, and plains, with a 200-mile river of hardened lava.
Moons	None.
Interesting factoids	Venus east to west compared with Earth's west to east rotation. Seen from Venus, the sun would rise in the west and set in the east. Atmospheric lightning bursts, long suspected by scientists, were finally confirmed in 2007 by the European *Venus Express* orbiter. On Earth, Jupiter, and Saturn, lightning is associated with water clouds, but on Venus, it is associated with clouds of sulfuric acid.

Mars

A belt of asteroids (fragments of rock and iron) between Mars and Jupiter separate the four inner planets from the five outer planets.

Distance from the Sun	141.71 million miles.
Distance from Earth	35 million miles.
Size	About one-half the size of Earth in diameter Volume: 0.151 x Earth Mass: 0.107 x Earth.
Gravity	If you weigh 100 pounds on Earth, you would weigh 38 pounds on Mars.
Length of year	687 Earth days.
Length of day	1.026 Earth days.
Surface temperature	From -125 °F to 23 °F.
Atmosphere	Carbon dioxide (95 percent).
First photographed	**1965**: NASA's *Mariner 4* sent back 22 photos of Mars, the world's first close-up photos of a planet beyond Earth.
First landings	**1976**: *Viking 1* and 2 landed on the surface of Mars. **1997**: *Mars Pathfinder* landed and dispatched *Sojourner*, the first wheeled rover to explore the surface of another planet. **2002**: *Mars Odyssey* began its mission to make global observations and find buried water ice on Mars. **2004**: Twin Mars Exploration Rovers named *Spirit* and *Opportunity* found the strongest evidence yet obtained that the Red Planet once had underground liquid water and water flowing on the surface.
Topology	Canyons, dunes, volcanoes, and polar caps of water ice and carbon dioxide ice.
Moons	2 Phobos and Deimos discovered in 1877.
Interesting factoids	Mars was named by the Romans for their god of war because of its red, blood-like color. Other civilizations also named this planet from this attribute; for example, the Egyptians named it "Her Desher," meaning "the red one." In 2008, the presence of water on Mars was confirmed.

Jupiter

Distance from the Sun	483.88 million miles.
Distance from Earth	At its closest, 370 million miles.
Size	11 times the diameter of Earth Diameter: 88,736 miles.
Gravity	If you weigh 100 pounds on Earth, you would weigh 265 pounds on Jupiter.
Length of year	12 Earth years.
Length of day	9 hours and 55 minutes.
Surface temperature	-234 °F.
Atmosphere	Whirling clouds of colored dust, hydrogen, helium, methane, water, and ammonia. The Great Red Spot is an intense windstorm larger than Earth.
First photographed	**1973:** *Pioneer 10* became the first spacecraft to cross the asteroid belt and fly past Jupiter.
First landings	As Jupiter is not solid it is not possible to actually land on the surface. **1995-2003:** The *Galileo* spacecraft dropped a probe into Jupiter's atmosphere and conducted extended observations of Jupiter and its moons and rings.
Topology	The composition of Jupiter's atmosphere is similar to that of the Sun – mostly hydrogen and helium. Deep in the atmosphere, the pressure and temperature increase, compressing the hydrogen gas into a liquid. At depths of about a third of the way down, the hydrogen becomes metallic and electrically conducting. At the center, the immense pressure may support a solid core of rock about the size of Earth.
Moons	63 moons and 4 rings. Io is the most volcanically active body in our Solar System. Ganymede is the largest planetary moon and the only moon in the Solar System known to have its own magnetic field. A liquid ocean may lie beneath the frozen crust of Europa, and icy oceans may also lie beneath the crusts of Callisto and Ganymede.
Interesting factoids	Jupiter is the largest planet in the Solar System. Its composition means that it is more like a star than a planet. Most of the visible clouds are composed of ammonia.

Saturn

■ **BELOW:** A black-and-white photo made by the *Cassini* in 2006 shows the swirling hurricane-like vortex at Saturn's south pole, where the vertical structure of the clouds is highlighted by shadows. Such a storm, with a well-developed eye ringed by towering clouds, is a phenomenon never before seen on another planet.

Distance from the Sun	887.14 million miles.
Distance from Earth	744 million miles at the closest point.
Size	About 10 times larger than Earth in diameter Diameter: 74,978 miles Volume: 763.594 x Earth Mass: 95.161 x Earth.
Gravity	If you weigh 100 pounds on Earth, you would weigh 107 pounds on Saturn.
Length of year	29.5 Earth years.
Length of day	10 hours, 40 minutes, 24 seconds.
Surface temperature	-288 °F.
Atmosphere	Hydrogen and helium.
First photographed	**1979:** *Pioneer 11* was the first spacecraft to reach Saturn, flying within 13,700 miles of the cloud tops.
First landings	No actual landings on Saturn as it is a gas giant but in **2005** the Huygens probe successfully landed on the moon Titan, returning images of the complex surface. **1981:** Using Saturn's powerful gravity as an interplanetary slingshot, *Voyager 2* was placed on a path toward Uranus.
Topology	Gas and liquid.
Moons	31 and over 1000 rings.
Interesting factoids	Since 2005 the *Cassini* spacecraft has returned high-resolution images of the Saturn system. Mission discoveries include evidence for liquid hydrocarbon lakes of methane and ethane on Titan, a new radiation belt around Saturn, new rings and moons, and icy jets and geysers at the south polar region of the moon Enceladus. Saturn's rings are flat and lie inside one another. They are made of billions of ice particles

Uranus

■ BELOW: These two pictures of Uranus were compiled from images recorded by *Voyager 2* on January 10, 1986, when the NASA spacecraft was 11 million miles from the planet. The view is toward the planet's pole of rotation, which lies just left of center. The picture on the left has been processed to show Uranus as the human eyes would see it from the vantage point of the spacecraft. The second picture is an exaggerated false-color view that reveals details not visible in the true-color view – including indication of what could be a polar haze of smog-like particles.

Distance from the Sun	1,783.98 million miles.
Distance from Earth	At the closest point, 1,607,000,000 miles.
Size	4 times larger than Earth in diameter Diameter: 32,193 miles Volume: 63.085 x Earth Mass: 14.536 x Earth.
Gravity	If you weigh 100 pounds on Earth, you would weigh 91 pounds on Uranus.
Length of year	84 Earth years.
Length of day	17 hours.
Surface temperature	Uniform temperature of -353°F.
Atmosphere	Hydrogen, helium, and methane. Uranus gets its blue-green color from methane gas in the atmosphere.
First photographed	Most of what we know about Uranus came from *Voyager 2*'s flyby in 1986. The spacecraft discovered 10 additional moons and several rings before heading on to Neptune.
First landings	N/A.
Topology	Ice giant.
Moons	27 moons and 11 rings.
Interesting factoids	Uranus wasn't discovered until 1781. Its discoverer, William Herschel, came from England. Like Venus, Uranus rotates east to west.

Neptune

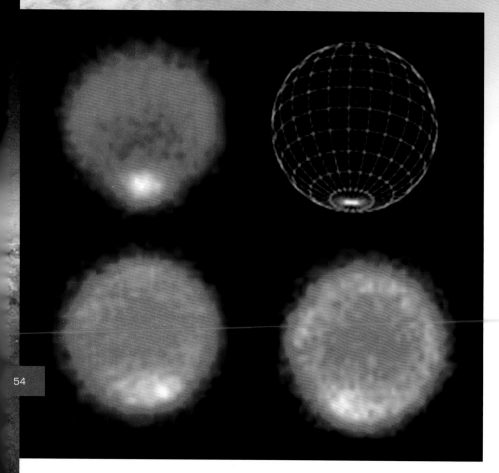

■ **LEFT:** This image provided by NASA shows thermal images of the planet Neptune. The images were obtained with the Very Large Telescope in Chile, using an imager/spectrometer for mid-infrared wavelengths on September 1 and 2, 2006. The upper left image samples temperatures near the top of Neptune's troposphere (near 100 millibar pressure, which is one-tenth the Earth atmospheric pressure at sea level). The hottest temperatures are indicated at the lower part of the image, at Neptune's south pole (see the graphic at the upper right). The lower two images, taken 6.3 hours apart, sample temperatures at higher altitudes in Neptune's stratosphere. They do show generally warmer temperatures near, but not at, the south pole. They also show a distinct warm area which can be seen in the lower left image and rotated completely around the back of the planet and returned to the Earth-facing hemisphere in the lower right image.

Distance from the Sun	2,796.46 million miles.
Distance from Earth	2,680,000,000 miles at closest point.
Size	Almost four times the size of Earth in diameter Diameter: 30,775 miles Volume: 57.723 x Earth Mass: 17.148 x Earth.
Gravity	If you weigh 100 pounds on Earth, you would weigh 114 pounds on Neptune.
Length of year	165 Earth years.
Length of day	16 hours and 7 minutes.
Surface temperature	-353°F.
Atmosphere	Hydrogen, helium, methane.
First photographed	**1989:** *Voyager 2* became the first and only spacecraft to visit Neptune, passing about 2,983 miles above the planet's north pole.
First landings	N/A.
Topology	A liquid layer covered with thick clouds and with constant, raging storms.
Moons	13 moons and 4 rings.
Interesting factoids	**1846:** Using mathematical calculations rather than through regular observations of the sky, astronomers discovered Neptune. It is invisible to the naked eye because of its extreme distance from Earth. Neptune's winds can be three times stronger than Jupiter and nine times stronger than Earth.

Pluto
– A Dwarf
Planet

What is a planet? That question has been asked since Greek astronomers came up with the word to describe the bright points of light that seemed to move among the fixed stars. Our Solar System's planet count has been as high as 15 before it was decided that some discoveries were different and should be called asteroids.

Many disagreed in 1930 that Pluto should be added as our Solar System's ninth planet. The debate flared again in 2005 when Eris – about the same size as Pluto – was found deep in a zone beyond Neptune called the Kuiper Belt. Was it the 10th planet? Or are Eris and Pluto examples of an intriguing, new kind of world?

The International Astronomical Union decided in 2006 that a new system of classification was needed to describe these new worlds, which are more developed than asteroids, but different than the known planets. Pluto, Eris, and the Asteroid Ceres became the first "dwarf planets." Unlike planets, dwarf planets lack the gravitational muscle to sweep up or scatter objects near their orbits. They end up orbiting the Sun in zones of similar objects such as the asteroid and Kuiper belts.

■ **BELOW: A 30-second exposure photograph shows attendees watching an image of Pluto with its moons, right, during** *To Pluto & Beyond!*, **a show at Barlow Planetarium on the University of Wisconsin-Fox Valley campus in Menasha, 2006. Pluto lost its status as the ninth planet of the Solar System earlier in the day, and has been reclassified by the International Astronomical Union as a "dwarf planet."**

"To **Boldly** Go"...
Space in Quotes

"I believe that this nation should commit itself to achieving the goal, before this decade is out, of landing a man on the Moon and returning him safely to the Earth. No single space project... will be more exciting, or more impressive to mankind, or more important... and none will be so difficult or expensive to accomplish."
President John F. Kennedy

"We go into space because whatever mankind must undertake, free men must fully share."
President John F. Kennedy

"There are so many benefits to be derived from space exploration and exploitation; why not take what seems to me the only chance of escaping what is otherwise the sure destruction of all that humanity has struggled to achieve for 50,000 years?"
Isaac Asimov

"I cannot join the space program and restart my life as an astronaut, but this opportunity to connect my abilities as an educator with my interests in history and space is a unique opportunity to fulfill my early fantasies."
Christa McAuliffe, teacher who died aboard *Challenger*, 1986

"To be the first to enter the cosmos, to engage, single-handed, in an unprecedented duel with nature – could one dream of anything more?"
Yuri Gagarin prior to flight, 1961

"Spaceflights cannot be stopped. This is not the work of any one man or even a group of men. It is a historical process which mankind is carrying out in accordance with the natural laws of human development."
Yuri Gagarin

"Earth is the cradle of humanity, but one cannot live in the cradle forever."
Konstantin Tsiolkovsky, Russian scientist, 1857

"The regret on our side is, they used to say years ago, we are reading about you in science class. Now they say, we are reading about you in history class."
Neil Armstrong, July 1999

"Many say exploration is part of our destiny, but it's actually our duty to future generations and their quest to ensure the survival of the human species."
Buzz Aldrin on the 37th anniversary of the *Apollo 11* landing, July 2006

"Two things are infinite: the universe and human stupidity; and I'm not sure about the universe."
Albert Einstein

"The inspirational value of the space program is probably of far greater importance to education than any input of dollars... A whole generation is growing up which has been attracted to the hard disciplines of science and engineering by the romance of space."
Arthur C. Clarke, author, 1970

"We can continue to try and clean up the gutters all over the world and spend all of our resources looking at just the dirty spots and trying to make them clean. Or we can lift our eyes up and look into the skies and move forward in an evolutionary way."
Buzz Aldrin, astronaut

"I don't know what you could say about a day in which you have seen four beautiful sunsets."
John Glenn, astronaut

"... to explore strange new worlds, to seek out new life and new civilizations, to boldly go where no one has gone before."
Gene Roddenberry, *Star Trek* creator

"Anyone who sits on top of the largest hydrogen-oxygen fueled system in the world, knowing they're going to light the bottom – and doesn't get a little worried – does not fully understand the situation."
John Young, astronaut

"Equipped with his five senses, man explores the universe around him and calls the adventure Science."
Edwin Hubble, scientist

The Legacy
of the
Space Race

■ ABOVE: The Space Shuttle *Atlantis*
is pictured prior to a perfect docking
with the International Space Station.
Part of a Russian Progress spacecraft
which is docked to the station is in the
foreground, 2011.

The Legacy of the Space Race

■ ABOVE: The Space Shuttle *Atlantis* is pictured prior to a perfect docking with the International Space Station. Part of a Russian Progress spacecraft which is docked to the station is in the foreground, 2011.

"The regret on our side is, they used to say years ago, we are reading about you in science class. Now they say, we are reading about you in history class."
Neil Armstrong, July 1999

"Many say exploration is part of our destiny, but it's actually our duty to future generations and their quest to ensure the survival of the human species."
Buzz Aldrin on the 37th anniversary of the *Apollo 11* landing, July 2006

"Two things are infinite: the universe and human stupidity; and I'm not sure about the universe."
Albert Einstein

"The inspirational value of the space program is probably of far greater importance to education than any input of dollars... A whole generation is growing up which has been attracted to the hard disciplines of science and engineering by the romance of space."
Arthur C. Clarke, author, 1970

"We can continue to try and clean up the gutters all over the world and spend all of our resources looking at just the dirty spots and trying to make them clean. Or we can lift our eyes up and look into the skies and move forward in an evolutionary way."
Buzz Aldrin, astronaut

"I don't know what you could say about a day in which you have seen four beautiful sunsets."
John Glenn, astronaut

"... to explore strange new worlds, to seek out new life and new civilizations, to boldly go where no one has gone before."
Gene Roddenberry, *Star Trek* creator

"Anyone who sits on top of the largest hydrogen-oxygen fueled system in the world, knowing they're going to light the bottom – and doesn't get a little worried – does not fully understand the situation."
John Young, astronaut

"Equipped with his five senses, man explores the universe around him and calls the adventure Science."
Edwin Hubble, scientist

■ **ABOVE:** Fireworks burst in the air over Space Shuttle *Atlantis* after it arrived at its new home at the Kennedy Space Center Visitor Complex, 2012, in Cape Canaveral. Accompanied by a fleet of astronauts spanning NASA's entire existence, *Atlantis* made a slow, solemn journey to retirement: the last space shuttle to orbit the world.

■ **BELOW:** Space Shuttle *Discovery* lifts off from pad 39A at the Kennedy Space Center, in Cape Canaveral, 2000.

The Shuttle

The Space Shuttle was designed to replace the single use rockets of the Apollo era with a reusable "space bus" and make space travel cheap, reliable, and everyday. Sadly, the shuttle was never as reusable as hoped. Maintaining the fleet took significant technical work and instead of flying once a week, the shuttle made on average fewer than five flights a year.

Although the shuttle ended up costing far more than intended and was subject to some high profile accidents – those that destroyed *Challenger* in 1986 and *Columbia* in 2003 killed 14 astronauts – we should not let the failures of the shuttle eclipse the program's achievements.

The shuttle enabled three of NASA's four "great observatories" to be put into orbit – the Hubble Space Telescope, the Compton Gamma Ray Observatory, and the Chandra X-ray Observatory. Together, these observatories give us an unprecedented view of the skies.

Several spacecraft were lifted into orbit on the shuttle before starting their onward journeys: the Galileo probe to photograph Jupiter, Magellan to map Venus, and the European Space Agency's Ulysses spacecraft to conduct the first survey of the Sun's environment.

Experiments carried out on the shuttle looked at how materials and living organisms behaved in the microgravity conditions of orbit. Did spiders spin webs, did seeds grow well, could fish still swim upright? (Answers: yes, sort of, no.) More valuable experiments revealed how weightlessness caused debilitating muscle and bone wastage on the human body. From this knowledge, medics have been able to draw up training programs to protect astronauts in orbit and possibly on lengthy missions to other planets.

But above all, the shuttle enabled the International Space Station, the world's first orbiting science lab.

The International Space Station

On November 2, 2010 the International Space Station (ISS) marked its 10th anniversary of continuous human occupation. In these 10 years the ISS has been visited by 204 individuals, traveled more than 1.5 billion miles (the equivalent of eight round trips to the Sun), and made 57,361 orbits around the Earth.

As of July 2012, there had been 125 launches to the ISS: 81 Russian vehicles, 37 space shuttles, one US commercial vehicle, three European, and three Japanese vehicles. A total of 162 spacewalks have been conducted in support of space-station assembly, totaling more than 1,021 hours.

The ISS, including its large solar arrays, spans the area of a football field, including the end zones, and weighs 861,804 pounds, not including visiting vehicles. The complex now has more livable room than a conventional five-bedroom house, and has two bathrooms, a gymnasium, and a 360-degree bay window.

An unprecedented international partnership of space agencies provides and operates the elements of the ISS. The principal countries are the United States, Russia, Europe, Japan, and Canada. The ISS has been the most politically complex space exploration program ever undertaken.

Scientifically it has provided a no-gravity environment for carrying out many different experiments, and in particular has allowed the effects of living in a low-gravity environment for prolonged periods to be studied in detail. If man is ever to explore deeper into space, then we need to know how to counteract any deleterious effects on the human body.

■ OPPOSITE: An artist's impression of the International Space Station.

■ BELOW: Astronaut Philippe Perrin, STS-111 mission specialist, floats near the Microgravity Science Glovebox (MSG) in the Destiny laboratory on the ISS.

Discoveries made on the ISS by international teams include:

- Purer protein crystals may be grown in space than on Earth. Analysis of these crystals helps scientists understand the fundamental building blocks of life. This type of research could lead to possible new treatments for cancer, diabetes, emphysema, and immune system disorders.

- Living cells can be grown in a laboratory environment in space where they are not distorted by gravity. Such cells can be used, for example, to test new treatments for cancer without risking harm to patients.

- Fluids, flames, molten metal, and other materials are the subject of basic research on the station. Scientists may be able to create better metal alloys and more perfect materials for applications such as computer chips.

- Some experiments take place on the exterior of the ISS. Exterior experiments study the space environment and how long-term exposure to space, vacuum, and debris affects materials. This research can provide future spacecraft designers a better understanding of the nature of space and so enhance spacecraft design. It may also lead to down-to-earth developments such as clocks a thousand times more accurate than today's atomic clocks, better weather forecasting, and stronger materials.

- Observations of the Earth from orbit help the study of large-scale, long-term changes in the environment. This can increase understanding of the forests, oceans, and mountains, the effects of volcanoes, ancient meteorite impacts, hurricanes, and typhoons.

Space Junk

Space junk is increasingly getting in the way of safe space travel – the International Space Station is always having to maneuver to avoid hitting lumps of metal flying around the Earth at hundreds of meters per second. The European Space Agency (ESA) believes there are currently 22,000 objects bigger than a coffee cup currently in orbit, of which only 1,100 are working satellites. What is more, there are more than 100 million objects smaller than one centimeter. This may not seem very big but, traveling at the speed of a bullet, they can still cause significant damage to a working satellite.

A number of organizations are looking at ways of clearing up this rubbish heap in the sky. One UK company is looking to develop a "garbage collector" satellite which would collect up space junk by harpooning it, then pushing or dragging it into the atmosphere where it would be burned up and disintegrate. NASA has proposed "zapping" space junk with a high-powered laser to disintegrate it as it passes. Boeing has filed a patent for sending up a satellite filled with heavy gases that, when sprayed on orbital debris, would slow them down enough to make them fall back into the atmosphere.

Space junk could be big business in future.

Voyager 1 and *2* – Beyond the Solar System

When *Voyager 1* and *2* finally leave the Solar System, what will they find? Or what will find them? In 2012 scientists discovered the nearest planet so far outside our Solar System, circling one of the stars of Alpha Centauri just four light-years away. The planet has at least the same mass as Earth, but circles its star far closer than Mercury orbits our Sun so is probably not habitable as it would be too hot. Even if the *Voyagers* keep going, it will take tens of thousands of years for them to arrive at the next Solar System.

The Space Race has fired our interest in science fiction, which has been fed through movies and TV shows such as *Star Trek, Star Wars,* and *Stargate.* But it will take a scientific leap here on Earth or by scientists in a distant Solar System to enable interstellar travel on the scale shown in these programs.

■ **ABOVE: A white arrow points to damage on a piece of a solar array from the Russian space station *Mir* in a display at the orbital debris program lab at Houston's Johnson Space Center, 2006. The array had been damaged by a miniscule piece of space junk.**

The Developing World and the Space Race

Is the balance of power in the Space Race shifting? Although China was relatively late to join in, it has started to launch shuttle-type craft which will enable the development of a large, modular space station that will probably be occupied on a semi-permanent basis. The Chinese space station will not approach the size or complexity of the ISS, but will be entirely built, launched, and controlled by China.

China is also taking advantage of using the world's existing knowledge of spaceflight to its advantage. Its first astronaut-carrying spacecraft was not a simple orbital capsule but a large, sophisticated vehicle. In addition to human spaceflight, China has sent two spacecraft to orbit the Moon, and has sent its latest Moon probe on an extended mission into deep space. It plans to land rovers on the Moon and retrieve samples. Robot missions to Mars and other targets in the Solar System are also proposed.

China is building an autonomous and sustainable space program. It controls its own rockets, satellites, crewed spacecraft, and space laboratories at a time when the major investor in the ISS (the US) can no longer send its own astronauts up there.

India, large but by no means as wealthy as China, is also investing in space flight. It already has capability to launch satellites and is looking at human space flight to further its research and development programs. Even Brazil has made investments in space development, teaming up with China and Russia in the process.

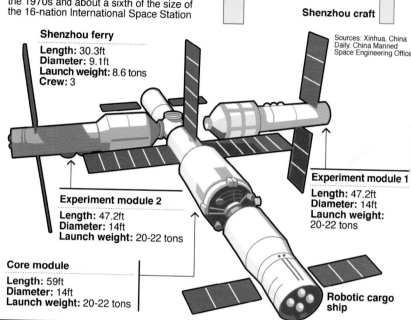

Tiangong (Heavenly palace): Chinese space station

Tiangong 1 is a prototype, and the plan is to eventually replace it with a permanent - and bigger - space station due for completion around 2020

Tiangong 1

Shenzhou craft

The future space station

The permanent station will weigh about 60 tons, slightly smaller than Nasa's Skylab of the 1970s and about a sixth of the size of the 16-nation International Space Station

Sources: Xinhua, China Daily, China Manned Space Engineering Office

Shenzhou ferry
Length: 30.3ft
Diameter: 9.1ft
Launch weight: 8.6 tons
Crew: 3

Experiment module 1
Length: 47.2ft
Diameter: 14ft
Launch weight: 20-22 tons

Experiment module 2
Length: 47.2ft
Diameter: 14ft
Launch weight: 20-22 tons

Core module
Length: 59ft
Diameter: 14ft
Launch weight: 20-22 tons

Robotic cargo ship

Press Association Graphic

■ **ABOVE:** The Chinese space station.

■ **BELOW:** Visitors to an exhibition about *Shenzhou 5*, China's first successful manned space mission, stand near models of the launch rockets in Beijing, China.